BAHAMAS PRIMARY

Mathematics Workbook 5

The authors and publishers would like to thank the following members of the Teachers' Panel, who have assisted in the planning, content and development of the books:

Chairperson: Dr Joan Rolle, Senior Education Officer, Primary School Mathematics, Department of Education

Team members:

Lelani Burrows, Anglican Education Authority

Deidre Cooper, Catholic Board of Education

LeAnna T. Deveaux-Miller, T.G. Glover Professional Development and Research School

Dr Marcella Elliott Ferguson, University of The Bahamas

Theresa McPhee, Education Officer, High School Mathematics, Department of Education

Joelynn Stubbs, C.W. Sawyer Primary School

Dyontaleé Turnquest Rolle, Eva Hilton Primary School

Karen Morrison, Rentia Pretorius and Lisa Greenstein

HODDER
EDUCATION
AN HACHETTE UK COMPANY

Hachette UK's policy is to use papers that are natural, renewable and recyclable products and made from wood grown in well-managed forests and other controlled sources. The logging and manufacturing processes are expected to conform to the environmental regulations of the country of origin.

Orders: Orders: please contact Hachette UK Distribution, Hely Hutchinson Centre, Milton Road, Didcot, Oxfordshire, OX11 7HH. Telephone: +44 (0)1235 827827. Email: education@hachette.co.uk. Lines are open from 9 a.m. to 5 p.m., Monday to Friday. You can also order through our website: www.hoddereducation.com

ISBN: **978 1 4718 6472 8**

© Cloud Publishing Services 2017

First published in 2017 by
Hodder Education,
An Hachette UK Company
Carmelite House
50 Victoria Embankment
London EC4Y 0DZ

www.hoddereducation.com

Impression number   10 9 8 7 6 5 4

Year   2023

Cover photo © Shutterstock/Worachat Sodsri

Illustrations by Peter Lubach and Aptara Inc.

Typeset in India by Aptara Inc.

Printed and bound by CPI Group (UK) Ltd, Croydon, CR0 4YY

A catalogue record for this title is available from the British Library.

Contents

Topic 1   Getting Ready   1

Topic 2   Numbers and Place Value   4

Topic 3   Exploring Patterns   6

Topic 4   Temperature   9

Topic 5   Fractions   13

Topic 6   Mass   16

Topic 7   Decimals   19

Topic 8   Classifying Shapes   22

Topic 9   Rounding and Estimating   27

Topic 10   Factors and Multiples   29

Topic 11   Lines and Angles   33

Topic 12   Number Facts   38

Topic 13   Adding and Subtracting   40

Topic 14   Statistics   42

Topic 15   Problem Solving   46

Topic 16   Length   49

Topic 17   Multiplying and Dividing   53

Topic 18   Order of Operations   56

Topic 19   Working with Time   58

Topic 20   Calculating with Fractions and Decimals   63

Topic 21   Capacity and Volume   65

Topic 22   Transformations   68

Topic 23   Perimeter and Area   71

Topic 24   Probability   73

Topic 1 Getting Ready

Check Your Skills

1 Complete these magic squares. You may not repeat any numbers in the same square.

8		4
	5	
6	7	2

7	0	5
3		1

4	9	
	5	
		6

2 Complete the table by rounding.

Number of People	To the Nearest Ten	To the Nearest Hundred	To the Nearest Thousand	To the Nearest Ten Thousand
28 458				
59 445				
28 999				
123 987				
148 755				

3 Make equivalent pairs.

A 0.4   **B** 0.5   **C** 1.5   **D** 3.75   **E** 0.02   **F** 0.1

a $\dfrac{2}{100}$ **b** $1\dfrac{1}{2}$ **c** $\dfrac{50}{100}$ **d** $\dfrac{1}{10}$ **e** $3\dfrac{3}{4}$ **f** $\dfrac{40}{100}$

4 Draw a bar graph to show the data in the table.

School	Number of Contestants in the Mathematics Quiz
A	40
B	25
C	50
D	30
E	25
F	45

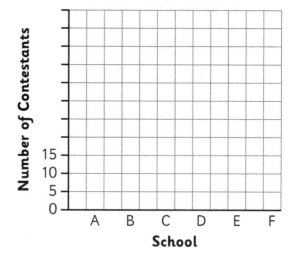

Answer these questions using the graph.

a What is the mode of the data? _____

b Which schools have the same number of contestants? _____

c How many contestants in all? _____

5 Draw lines of symmetry on each shape. Tick the shapes that are polygons.

a

b

c

d

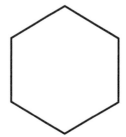

6 Write odd and even numbers on each spinner to show the probability of landing on an odd number to be:

a impossible

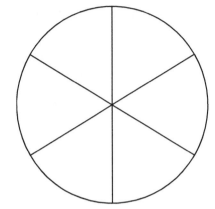

b certain

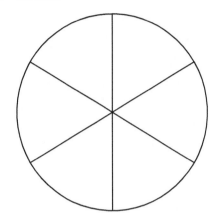

c one in six chance

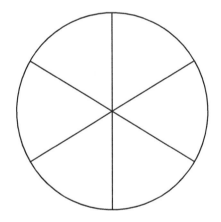

d one in three chance

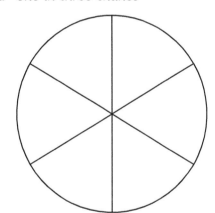

7 Draw a new shape to show each transformation.

a

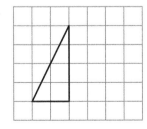

Flip

b

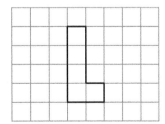

Slide two bocks left

c

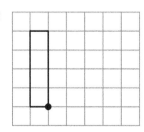

Make a $\frac{1}{4}$ turn clockwise

Topic 2 Numbers and Place Value

Revisiting Millions

1 Sort the numbers in the box into two groups. Write them in the correct columns in the table.

4 967 204	55 984 309	66 986	890	54 002
78 008	898 765	4 321 000	999	876 987
999 999	9 999 999	34 210 342	654 329	1 000 001

> 1 Million	< 1 Million

2 Write the missing numbers on the number line.

a

4 000 000   6 000 000   9 000 000

b

2 500 000   3 500 000

c

9 700 000   9 750 000

More Millions

1 Fill in < or > to make each statement true.

a 34 909 123 ☐ 34 912 345

b 120 341 256 ☐ 102 789 966

c 499 459 093 ☐ 498 098 432

d 315 090 450 ☐ 315 090 540

2 Use data from the table on page 9 Question 6 of your Student Book to label this graph correctly.

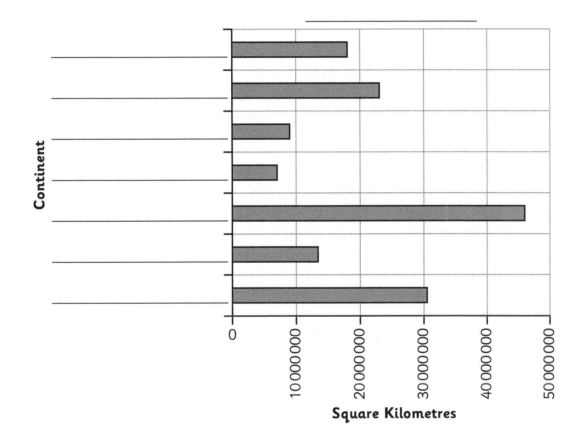

3 Complete these sentences about the graph.

a The graph shows the _____

b The three largest _____

c Asia is almost double the size of _____

Topic 3 Exploring Patterns

Patterns Around Us

Find six patterns in your local area. They can be natural or made by people.

Draw the patterns.

Write where you found each pattern.

Matching Patterns

1 Draw the next shape in each geometrical pattern in column 1.

2 Match the geometrical patterns to the number patterns in column 2.

Column 1	Column 2
a	3, 5, 7, 9, 11 …
b	1, 4, 7, 10, 13 …
c	2, 4, 6, 8, 10 …
d	1, 3, 5, 7, 9 …
e	3, 6, 9, 12, 15 …
f	1, 4, 7, 10, 13 …

Describe and Extend Patterns

1 Find the rule then use the rule to work out the next 3 numbers in each pattern.

a 11, 13, 15, 17, _____, _____, _____

b 21, 30, 39, 48, _____, _____, _____

c 2, 4, 8, 16, _____, _____, _____

d 1, 5, 25, _____, _____, _____

e 2, 6, 18, 54, _____, _____, _____

f 3, 9, 27, _____, _____, _____

g 0.001, 0.01, 0.1, 1, _____, _____, _____

h 1 000, 500, 250, _____, _____, _____

2 What are the missing numbers in these patterns? Compare your answers with a partner.
Tell your partner how you worked them out.

a 1, 3, 9, _____, _____, 243

b 59, 54, _____, _____, 39, _____

c 1, 2, 4, _____, _____, 32, _____

d 157, _____, _____, 127, 117

e 81, 27, _____, _____, 1, _____

f 93, _____, 77, 69, 61, _____, _____

g 122, 223, 324, _____, _____, _____

h 176, 88, 44, _____, _____, _____

Topic 4 Temperature

Working with Temperature

1 Circle the word that fits best in each sentence.

    a The temperature outside today is **warm / cold / icy / hot**.

    b The units we use to measure temperature are **metres / grams / degrees**.

    c 100 °C is the **freezing point / boiling point** of water.

    d When we make something colder, we **lower / raise / improve** the temperature.

    e When we make something hotter, we **bring down / bring up / bring on** the temperature.

2 These thermometers show temperatures in degrees Celsius and Fahrenheit.
Read the temperatures on each thermometer, and write them in the
answer spaces.

a

°C	°F
50	122
55	131
60	140
65	149
70	158
75	167
80	176
85	185
90	194
95	203
100	212

b

290°C

554°F

c

°C	°F
100	220
90	200
80	180
70	160
60	140
50	120
40	100
30	80
20	60
10	40
0	20
−10	0

d

_____   _____   _____   _____

3 On the line provided, write the letter to match each thermometer reading to the correct picture. Use the table on page 21 in your Student Book to help you.

a

Hot oven _____

b

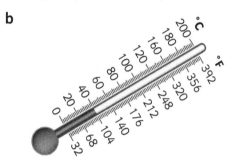

Freezing point of water _____

c

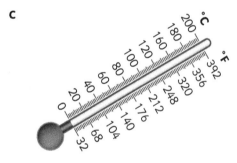

Boiling water _____

d

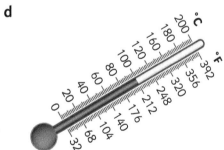

Hot bath _____

e

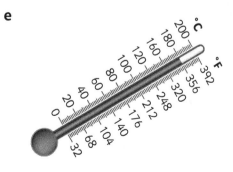

Cup of tea _____

4 The table below shows the temperature in some different cities in the world on a day in December.

Complete the table by converting between units. You can use a calculator to help you.

Place	Temperature in °C	Temperature in °F
London, England	7	
Nairobi, Kenya	18	
Perth, Australia		86
Rio de Janeiro, Brazil	26	
Helsinki, Finland	−4	
Paris, France	4	

5 Now answer some questions about the table above.

a The coldest city was _____.

b The warmest city was _____.

c _____ was warmer than Ontario but cooler than London.

d It was _____ degrees hotter in Perth than in Rio.

e It was _____ degrees colder in London than in Nairobi.

6 Circle the warmer temperature in each pair.

a	0 °C	25 °F
b	18 °C	50 °F
c	100 °C	100 °F
d	12 °C	0 °F
e	140 °F	50 °C
f	100 °F	60 °C

Working with Negative Temperatures

This table shows you the lowest recorded temperatures from some of the world's coldest places.

Place	Lowest Recorded Temperature (°F)
Rogers Pass – Montana, USA	–70
Vostok – Antarctica	–128.6
Fort Selkirk – Yukon, Canada	–74
Prospect Creek – Alaska, USA	–80
Oymyakon – Russia	–96.2
Elsmitte – Greenland	–85

a   Which of these places has the coldest recorded temperature?

_____

b   What is the difference between the temperature for Elsmitte and the temperature for Prospect Creek?

_____

c   How much colder was the temperature in Fort Selkirk than in Rogers Pass?

_____

d   Work out the difference between the temperatures for Oymyakon and Elsmitte.

_____

e   Look at the temperature for Prospect Creek. What would the temperature be if it dropped another 3 degrees?

_____

Topic 5 Fractions

Revisiting Fractions

1 Write the letter to match the fractions on the left to the equivalent fraction on the right.

2 Circle all the fractions that are in simplest form.

a $\frac{15}{20}$

b $\frac{4}{10}$

c $\frac{16}{20}$

d $\frac{9}{12}$

e $\frac{25}{100}$

f $\frac{1}{2}$

g $\frac{2}{3}$

h $\frac{8}{14}$

i $\frac{3}{4}$

j $\frac{7}{21}$

A $\frac{2}{5}$

B $\frac{14}{28}$

C $\frac{1}{4}$

D $\frac{1}{3}$

E $\frac{4}{7}$

F $\frac{8}{10}$

G $\frac{6}{8}$

H $\frac{75}{100}$

I $\frac{16}{24}$

J $\frac{3}{4}$

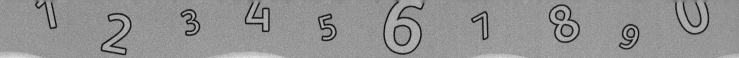

3 Complete the statements to make them true.

a $\dfrac{1}{2} = \dfrac{\square}{6}$

b $\dfrac{3}{4} = \dfrac{\square}{12}$

c $\dfrac{3}{5} = \dfrac{\square}{10}$

d $\dfrac{7}{8} = \dfrac{\square}{16}$

e $\dfrac{7}{12} = \dfrac{\square}{48}$

f $\dfrac{\square}{100} = \dfrac{3}{5}$

g $\dfrac{5}{8} = \dfrac{15}{\square}$

h $\dfrac{4}{\square} = \dfrac{2}{3}$

i $\dfrac{20}{\square} = \dfrac{5}{12}$

4 Colour fractions of the circle according to the instructions.

a $\dfrac{1}{4}$ blue

b $\dfrac{1}{6}$ red

c $\dfrac{1}{3}$ green

d $\dfrac{1}{12}$ yellow

e $\dfrac{1}{9}$ pink

f $\dfrac{1}{18}$ brown

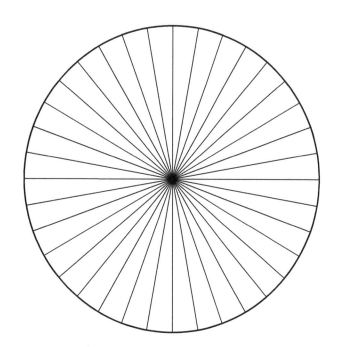

Compare and Order Fractions

1 Fill in <, = or > to make each statement true.

a $\dfrac{1}{5}$ ☐ $\dfrac{1}{2}$

b $\dfrac{1}{3}$ ☐ $\dfrac{1}{6}$

c $\dfrac{1}{2}$ ☐ $\dfrac{2}{4}$

d $\dfrac{2}{3}$ ☐ $\dfrac{2}{5}$

e $\dfrac{4}{7}$ ☐ $\dfrac{4}{8}$

f $\dfrac{2}{4}$ ☐ $\dfrac{6}{12}$

g $\dfrac{4}{7}$ ☐ $\dfrac{5}{9}$

h $\dfrac{14}{15}$ ☐ $\dfrac{9}{10}$

i $\dfrac{7}{8}$ ☐ $\dfrac{9}{27}$

j $\dfrac{3}{14}$ ☐ $\dfrac{6}{21}$

k $\dfrac{8}{9}$ ☐ $\dfrac{11}{12}$

l $\dfrac{5}{7}$ ☐ $\dfrac{3}{5}$

2 Write the fractions and mixed numbers in the correct positions on the number line.

$\dfrac{1}{2}$ $1\dfrac{7}{8}$ $2\dfrac{1}{4}$ $3\dfrac{4}{16}$ $4\dfrac{3}{24}$ $2\dfrac{1}{2}$ $\dfrac{14}{16}$

0   1   2   3   4   5

3 Write each fraction in the correct position on the number line.

$\dfrac{3}{9}$ $\dfrac{1}{4}$ $\dfrac{2}{3}$ $\dfrac{7}{12}$ $\dfrac{18}{24}$ $\dfrac{11}{12}$ $\dfrac{3}{3}$

0   1

Topic 6 Mass

Getting Started

1 a Use this table to record the items you compared to a crayon and a textbook.

Items That Have About the Same Mass as a Crayon (5 g)	Items That Have About the Same Mass as a Textbook (1 kg)

b Something lighter than a crayon (less than 5 g): _____

c Something heavier than a textbook (more than 1 kg): _____

2 Estimate the mass of each item in the table. Then use a scale to check your measurement.

How close was your estimate to the actual mass? Write the difference between your estimate and the actual mass.

Item	Estimate	Actual Mass	Difference
Homework diary			
Ruler			
Stapler			
Glue stick			
Pen			
Scissors			
Eraser			

Standard Units of Mass

1 Choose the most suitable mass for each item, and write it under the picture.

137 g	58 g	2.5 kg	12 g	500 g	1 t

a Bag of flour

_____

b Loaf of bread

_____

c Pile of bricks

_____

d Tennis ball

_____

e Key

_____

f Smartphone

_____

2 Find ten items in your home. Estimate the mass of each item in g or kg. Then use a scale to measure the mass.

Item	My Estimate	My Measurement

Converting Between Units

1 Complete the relationships between these units.

   a 1 kilogram (kg) = _____ grams (g)   **b** 2 kg = _____ g

   c 3 kg = _____ g   **d** 5 kg = _____ g

   e 0 kg = _____ g

2 The rule for changing from kg to grams is:

_____

3 Complete the relationships between these units:

   a 1 000 g = _____ kg   **b** 2 000 g = _____ kg

   c 10 000 g = _____ kg   **d** 1 500 g = _____ kg

4 Convert between g and kg to complete the tables. The first two have been filled in for you as examples.

Mass in Grams	Mass in Kilograms
750 g	0.75 kg
1 150 g	1.15 kg
225 g	
	2.95 kg
489 g	
1 565 g	

Mass in Grams	Mass in Kilograms
	3.188 kg
	7.335 kg
10 000 g	
12 055 g	
	11.5 kg
	18.35 kg

5 Solve these word problems.

   a A builder's pickup truck can carry a load of up to half a ton at a time.
   The builder needs to transport 2.6 tons of bricks to a site.

   How many trips must she make? _____

   b A tube of ointment has a mass of 20 g. The pharmacy receives a delivery of tubes.
   The total mass of the parcel is 705 g. The packaging materials have a mass of 225 g.

   How many tubes are in the parcel? _____

Topic 7 Decimals

Revisit Decimal Place Value

1 In the diagrams below, the cube contains 1 000 small cubes and represents 1 whole. Write the letter that matches each diagram to a decimal fraction.

a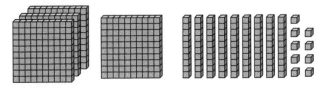

1.123 _____

b

0.499 _____

c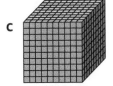

1.151 _____

d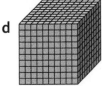

0.059 _____

e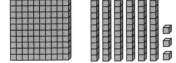

1.063 _____

2 Circle the digit 7 in each number. Write its value next to it.

a 3.072 _____   b 47.345 _____

c 0.782 _____   d 3.427 _____

e 70.005 _____   f 1.725 _____

Compare and Order Decimals

1 Fill in <, = or > to make these statements true.

a 0.812 ☐ 0.878

b 13.8 ☐ 12.8

c 0.876 ☐ 1.854

d 0.7 ☐ 0.07

e 0.30 ☐ 3.00

f 12.015 ☐ 12.055

g 0.987 ☐ 1

h 0.014 ☐ 0.023

i 0.298 ☐ 0.28

j 2.076 ☐ 2.06

k 1.852 ☐ 1.850

l 2.705 ☐ 2.750

2 Rewrite each set of decimals in ascending order

a 0.945   0.95   0.594   _____, _____, _____

b 3.133   2.131   8.033   _____, _____, _____

c 0.341   0.431   0.411   _____, _____, _____

d 0.050   0.5   0.550   _____, _____, _____

3 $\frac{1}{100}$ of a litre is added to each container. Write the new volume of water.

a

2.5 L

b

0.85 L

c

10.939 L

d

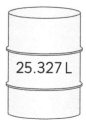

25.327 L

_____ L   _____ L   _____ L   _____ L

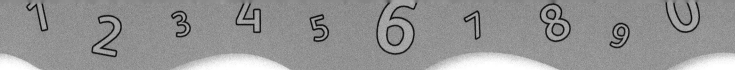

Convert Between Fractions and Decimals

Find your way through the mathematics maze.

Read the statement.

If it is true, follow the T arrow. If it is false, follow the F arrow.

START

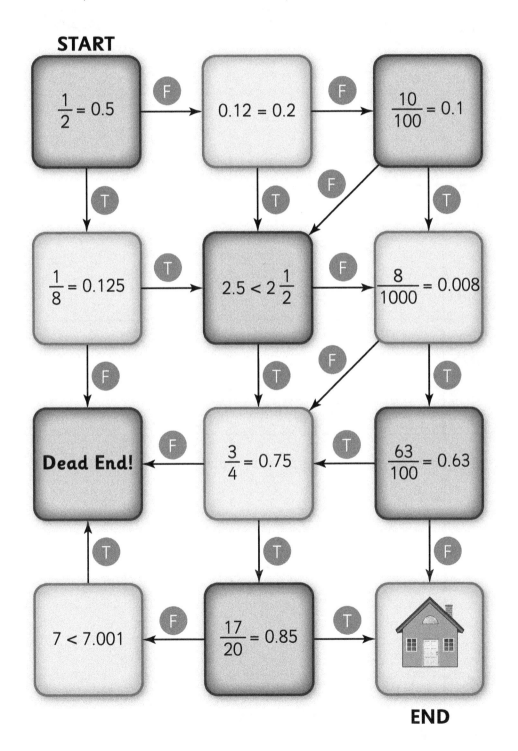

END

Topic 8 Classifying Shapes

Polygons

1 Tick in the columns that apply to each shape.

Curve		Open	Closed	Simple	Not Simple
a					
b					
c					
d					
e					

2 Circle all the regular polygons.

a 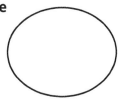 b  c 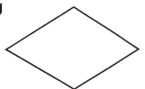 d

e 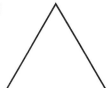 f 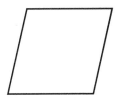 g  h

i 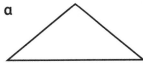 j 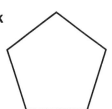 k   l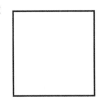

3 Draw a line from the name of the polygon on the left to the number of sides and angles it has on the right.

Hexagon	3
Pentagon	4
Nonagon	5
Quadrilateral	6
Heptagon	7
Decagon	8
Octagon	9
Triangle	10

Quadrilaterals

Read these statements.

A One set of opposite angles are equal.

B All opposite angles are equal.

C All opposite sides are the same length.

D There are no parallel sides.

E All angles are right angles.

F All opposite sides are parallel.

G The adjoining sides are the same length.

H All four sides are the same length.

I One set of opposite sides are parallel.

Write the letters of all the statements that apply to each quadrilateral inside it.

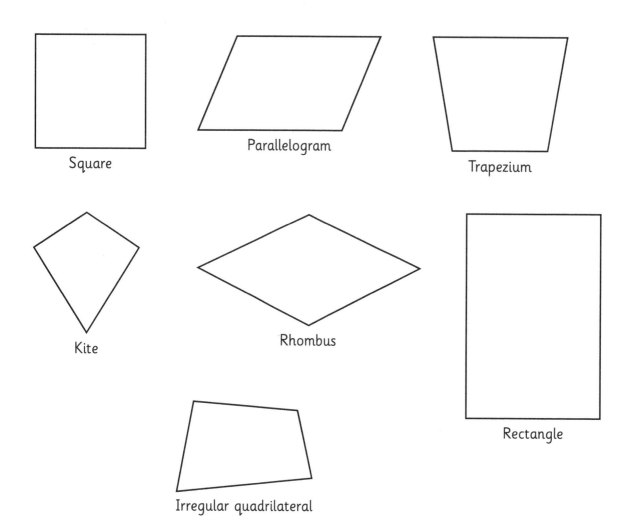

Square

Parallelogram

Trapezium

Kite

Rhombus

Rectangle

Irregular quadrilateral

Solids

1 Complete the table.

Solid	Name	Number of Faces	Number of Edges	Number of Vertices	Shape of Faces
a					
b					
c					
d					
e					
f					
g					
h					

2 For each solid, shade the correct pattern that would make the solid shape when folded.

a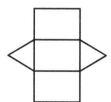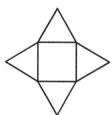

b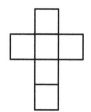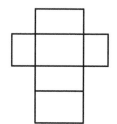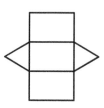

c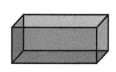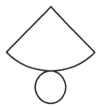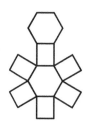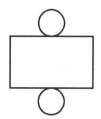

d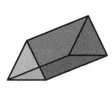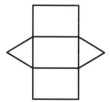

e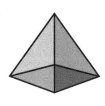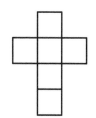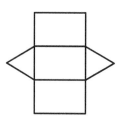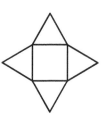

f

Topic 9 Rounding and Estimating

Round Whole Numbers and Decimals

1 Round the numbers to the place given in each column.

	Number	To Nearest Hundred	To Nearest Ten Thousand	To Nearest Million
a	3 209 876			
b	12 987 452			
c	9 099 100			
d	32 987 345			
e	12 987 555			
f	4 919 326			
g	5 245 765			

2 The decimals on the left have been rounded to the nearest tenth. Circle the decimals on the right that would give that answer when rounded.

a	0.400	0.276	0.352	0.389	0.399	0.349	0.427
b	0.700	0.642	0.654	0.701	1.073	0.756	0.627
c	1.20	1.19	1.24	1.21	1.476	1.32	1.09
d	3.50	3.71	3.05	3.45	3.54	3.62	3.09
e	9.00	8.79	9.29	9.02	8.55	8.95	9.99
f	10.00	10.09	9.98	9.99	10.45	9.47	9.95

Estimate Answers

1 Round the prices to the nearest dollar and estimate the total for each bill.

12.99	125.25	9.99
3.50	14.99	10.50
4.25	100.00	13.25
0.75	199.99	16.80
0.50	145.25	12.40
1.24		3.45
3.25		9.28

2 Toniqua used a calculator to do these calculations. Write an estimate below each calculation and tick the answers that you think are reasonable.

a 345 + 450 + 399 + 289

Estimate _____

b 312 + 689 + 5 234 + 459

Estimate _____

c 23 897 – 12 988

Estimate _____

d 43 126 – 24 999

Estimate _____

e 4 532 × 9

Estimate _____

f 1 278 ÷ 3

Estimate _____

g 199 × 9

Estimate _____

Topic 10 Factors and Multiples

1 Complete the multiples patterns.

a 8, 16, _____, _____, _____, _____, _____, _____, _____

b 7, 14, 21, _____, _____, _____, _____, _____, _____, _____

c 12, 24, _____, _____, _____, _____, _____, _____, _____

d 12, 15, 18, _____, _____, _____, _____, _____, _____, _____

e 42, 44, 46, _____, _____, _____, _____, _____, _____, _____

f 50, 100, 150, _____, _____, _____, _____, _____, _____, _____

2 One factor for each product is given. Complete the table by writing the other factor in each pair.

Product	16	16	16	18	18	18	24	24	24	24
Factor	1	2	4	6	9	18	24	12	4	8
Factor										

3 Use the information in the table to list all the factors of:

a 16 _____

b 18 _____

c 24 _____

4 List all the factors of each of these numbers.

a Factors of 36: _____, _____, _____, _____, _____, _____, _____, _____, _____

b Factors of 50: _____, _____, _____, _____, _____, _____

c Factors of 64: _____, _____, _____, _____, _____, _____, _____

5 Why do 36 and 64 have an odd number of factors? _____

_____

Greatest Common Factors and Least Common Multiples

1 Complete the tables. Write the greatest common factor.

Product	Factors
24	
36	

The GCF of 24 and 36 is _____

Product	Factors
32	
40	

The GCF of 32 and 40 is _____

Product	Factors
42	
70	

The GCF of 42 and 70 is _____

Product	Factors
30	
75	

The GCF of 30 and 75 is _____

2 List the multiples of each number. Find the LCM.

a 7 _____, _____, _____, _____, _____, _____

5 _____, _____, _____, _____, _____, _____, _____, _____

The LCM of 7 and 5 is _____

b 10 _____, _____, _____, _____, _____, _____, _____

12 _____, _____, _____, _____, _____, _____, _____

The LCM of 10 and 12 is _____

c 3 _____, _____, _____, _____, _____, _____, _____, _____

8 _____, _____, _____, _____, _____

The LCM of 3 and 8 is _____

Prime and Composite Numbers

Eratosthenes was a Greek mathematician who worked out a method for sorting prime numbers from composite numbers. This method is now called the sieve of Eratosthenes.

1 Follow the instructions to see how this works.

1	2	3	4	5	6	7	8	9	10
11	12	13	14	15	16	17	18	19	20
21	22	23	24	25	26	27	28	29	30
31	32	33	34	35	36	37	38	39	40
41	42	43	44	45	46	47	48	49	50
51	52	53	54	55	56	57	58	59	60
61	62	63	64	65	66	67	68	69	70
71	72	73	74	75	76	77	78	79	80
81	82	83	84	85	86	87	88	89	90
91	92	93	94	95	96	97	98	99	100

- Cross out 1.
- Circle 2 then cross out all the other multiples of 2.
- Circle 3 then cross out all the other multiples of 3.
- Circle the next number and then cross out all the multiples of this number.
- Repeat this until the table is complete.

2 List all the prime numbers from 1 to 100. (You should have 25)

———, ———, ———, ———, ———, ———, ———, ———, ———, ———,

———, ———, ———, ———, ———, ———, ———, ———, ———, ———,

———, ———, ———, ———, ———

3 How many composite numbers are there from 1 to 100? _____

4 We can write 16 as a sum of two prime numbers like this: 16 = 5 + 11.
Try to write these composite numbers as a sum of two prime numbers.

30 = _____ + _____   24 = _____ + _____   17 = _____ + _____   25 = _____ + _____

Prime Factors

Complete the factor trees. The circles may only contain prime factors.
Write each number as a product of its prime factors.

a

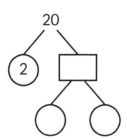

_____

b

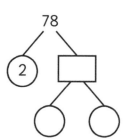

_____

c

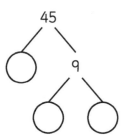

_____

d

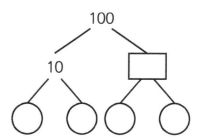

_____

e

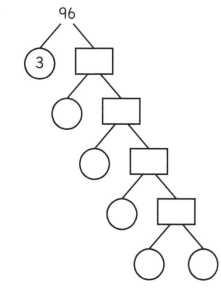

_____

f

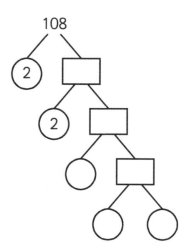

_____

Topic 11 Lines and Angles

Points, Lines and Planes

1 Look at the letters of the alphabet.

A B C D E F G H I J K L M N O P Q R S T U V W X Y Z

a How many line segments are in the letter F?   _____

How many angles are in the letter F?   _____

What type of angles are in the letter F?   _____

b Which letters have perpendicular line segments?

_____

c Which letters have parallel line segments?

_____

d Are there any letters that have perpendicular and parallel line segments? Name them.

_____

e Are there any letters that have no line segments? Name them.

_____

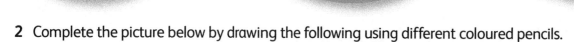

2 Complete the picture below by drawing the following using different coloured pencils.

- Line segment AB
- Ray EF
- Line segment CD
- Parallel line segments GA and HC
- Plane IJK
- Point L
- Perpendicular line segments MN and OP
- Point Q
- Parallel lines RS and TU

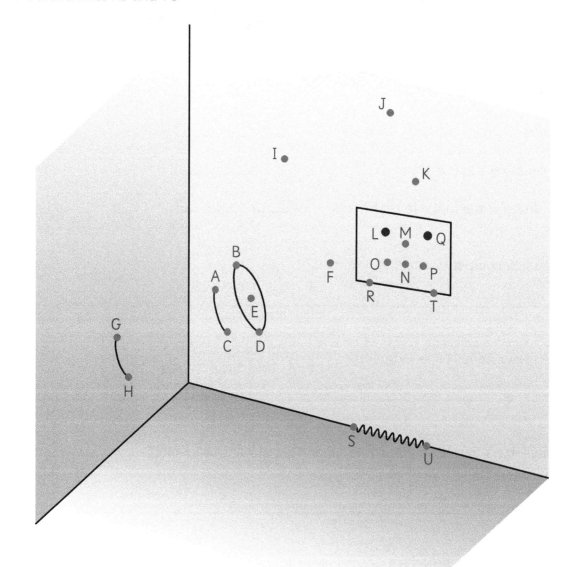

Angles

1 Draw an example of each type of angle.

Obtuse Angle	**Right Angle**
Full Rotation	**Straight Angle**
Acute Angle	**An Angle Bigger than 180°**

2 Draw hands on the clocks to show each type of angle.

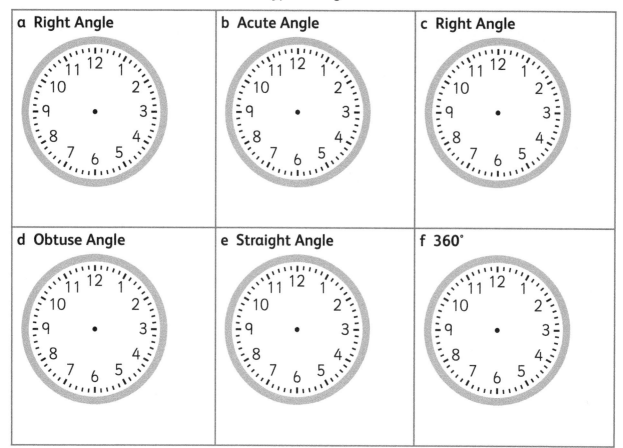

| a Right Angle | b Acute Angle | c Right Angle |
| d Obtuse Angle | e Straight Angle | f 360° |

Measuring Angles

3 Choose the correct size angle from the box for each of the angles below and write the size on the line next to the diagram.

| 90° | 45° | 110° | 75° | 360° | 160° | 15° |

a _____   b _____   c _____

d _____   e _____   f _____   g _____

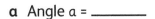

4 Use a protractor to measure the angles.

a Angle a = _____

  Angle b = _____

  Angle a + Angle b = _____

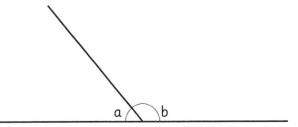

b Angle c = _____

  Angle d = _____

  Angle c + Angle d = _____

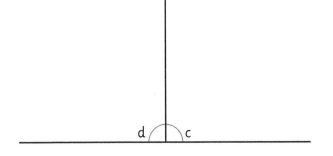

c Without measuring the angle, what is the size of angle ƒ? _____
  How do you know this?

  _____

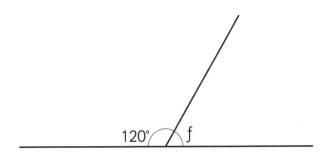

120°   ƒ

5 Use your protractor to answer the following questions.

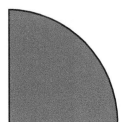

  a How many degrees is the rotation
    of the quarter circle on the right? _____
  b If you put two quarter circles together, how
    many degrees is the rotation of the angle? _____
  c Without using a protractor, how many degrees
    is the rotation of the angle if you put two half circles together?

  _____

Topic 12 Number Facts
Test Yourself

Write the answers. Record how long it takes you. Check your answers with a calculator and write your score.

Test 1	Test 2	Test 3
a 6 + 8 =	**a** 11 + 8 =	**a** 7 + 9 =
b 8 × 8 =	**b** 6 × 8 =	**b** 4 × 8 =
c 14 + 9 =	**c** 12 + 9 =	**c** 11 + 9 =
d 3 × 10 =	**d** 8 × 10 =	**d** 9 × 10 =
e 15 − 8 =	**e** 19 − 8 =	**e** 11 − 8 =
f 7 + 5 =	**f** 17 + 5 =	**f** 17 − 8 =
g 42 ÷ 6 =	**g** 48 ÷ 6 =	**g** 42 ÷ 3 =
h 3 × 7 =	**h** 8 × 7 =	**h** 3 × 9 =
i 5 × 9 =	**i** 6 × 9 =	**i** 5 × 4 =
j 81 ÷ 9 =	**j** 63 ÷ 9 =	**j** 30 ÷ 10 =
k 10 × 10 =	**k** 8 × 8 =	**k** 5 × 5 =
l 28 ÷ 4 =	**l** 28 ÷ 3 =	**l** 18 ÷ 9 =
m double 9 =	**m** double 11 =	**m** double 7 =
n half of 24 =	**n** half of 18 =	**n** half of 30 =
o 24 − 9 =	**o** 24 − 13 =	**o** 14 − 10 =
p 13 + 7 =	**p** 13 + 8 =	**p** 33 + 7 =
q 14 − 8 =	**q** 14 − 9 =	**q** 15 − 8 =
r 19 − 0 =	**r** 23 − 0 =	**r** 18 − 0 =
s 19 × 0 =	**s** 12 × 0 =	**s** 4 × 0 =
t 19 + 0 =	**t** 17 + 0 =	**t** 100 + 0 =
Time taken:	Time taken:	Time taken:
_____ Score $\frac{\square}{20}$	_____ Score $\frac{\square}{20}$	_____ Score $\frac{\square}{20}$

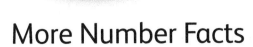

More Number Facts

Complete the tables as quickly and as accurately as you can.

a

Number	3	6	5	7	8	1	2	0	4	9
× 4										
× 8										
× 7										
× 3										
× 10										

b

Number	3	6	5	7	8	1	2	0	4	9
+ 9										
+ 8										
+ 2										
+ 5										
+ 10										

c

Number	13	16	15	17	18	11	12	10	14	19
− 9										
− 8										
− 2										
− 5										
−10										

d

Number	20	12	4	16	36	28	40	32	24	8
÷ 2										
÷ 4										
÷ 3	■		■	■			■	■		■
÷ 6	■		■	■		■	■			■
÷ 8	■	■	■		■	■				

Topic 13 Adding and Subtracting

Estimate and Calculate

1 Which operation sign is missing from each calculation?

    a 12 345 ☐ 13 346 = 25 691   **b** 123 000 ☐ 8 999 = 114 001

    c 65 412 ☐ 12 439 = 52 973   **d** 12 398 ☐ 14 899 = 27 297

2 Complete the table. Work on scrap paper if you need more space.

Calculation	Round Figures to Nearest Thousand	Estimated Answer	Actual Answer
4 568 + 5 412			
12 765 + 32 123			
18 999 + 17 544			
431 123 + 423 098			
32 233 – 19 233			
28 686 – 27 544			
12 456 – 9 656			
63 412 – 45 987			

Missing Values

The diagrams show the journey of different cruise ships. Distances are in kilometres.
Work out the missing distances for each journey and write them in the box provided.

a

5 627 km   4 837 km

b

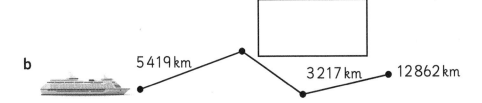

5 419 km   3 217 km   12 862 km

c

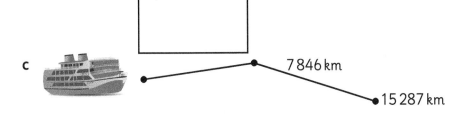

7 846 km

15 287 km

d

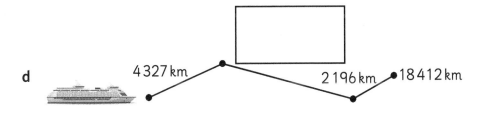

4 327 km   2 196 km   18 412 km

e

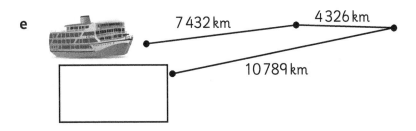

7 432 km   4 326 km

10 789 km

Topic 14 Statistics

Mean, Median, Mode and Range

1 Ella rolled a die a number of times and drew a graph of her results.

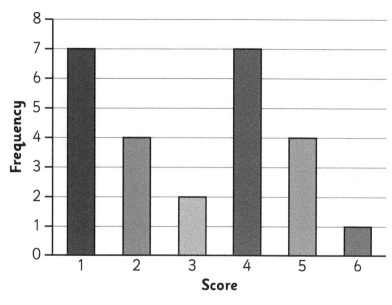

a List the results as a data set in numerical order.

_____

_____

b Calculate the mean of the data set.

$$\text{Mean} = \frac{\text{Sum of all the values}}{\text{Number of values}}$$

c What was the median score? _____

d What was the modal score? _____

2 An ice-cream shop has to give up one of its flavours. In order to determine which flavour was the worst seller, the owner gathered the following sales data over the period of a month:

	Week 1	Week 2	Week 3	Week 4
Carrot	3	3	25	2
Cucumber	8	8	7	5

a Calculate the mean for the sales of each flavour in order to find out which ice-cream flavour was the worst seller.

Carrot: _____

Cucumber: _____

b Look at the data entries. Do you think that using the mean gives a realistic view of the least favourite ice-cream flavour? Give a reason for your answer.

_____

c Now calculate the median and the mode for the sales of each flavour.

Carrot median: _____

Carrot mode: _____

Cucumber median: _____

Cucumber mode: _____

d What do the median and mode tell you about the popularity of carrot ice cream and cucumber ice cream?

_____

e What would you advise the owner of the ice-cream shop and what reasons would you give?

_____

_____

Analysing and Representing Data

Draw a graph to show the following events in order on the grid below.

- Byron walked for half an hour to the library that is three kilometres from his home.
- He studied at the library for one hour.
- He then walked to a friend's house. It took him one hour to walk 6 km at 6 km/h.
- He visited his friend for two hours.
- It took him half an hour to walk back to his house, which is one kilometre from his friend's house.

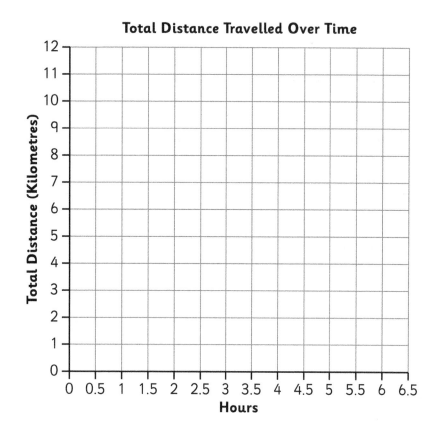

a How far did Byron walk in total?

_____

b Did he spend longer walking or not walking?

_____

c Did he always walk at the same speed?

_____

Venn Diagrams

Read the passage below, then complete the Venn diagram using summarized information from the passage. Enter at least three facts into each part of the diagram.

Butterflies and moths are flying insects with patterned wings. They each have six legs and four wings. Both butterflies and moths go through four stages of metamorphosis: from an egg to a larva, then a pupa and finally the adult butterfly or moth that we see flying around.

While butterflies are active during the day, moths are mostly nocturnal animals. Moths have plump bodies and feathery antennae. When they rest, their wings are down and spread over their bodies. Butterflies have slender bodies with thin, straight antennae. They hold their wings together above their bodies when they are at rest.

Characteristics of Moths and Butterflies

Moths

Butterflies

Topic 15 Problem Solving

Non-Routine Problems

1 Fill in the digits from 1 to 7 so that each line of three hexagons has the same sum.

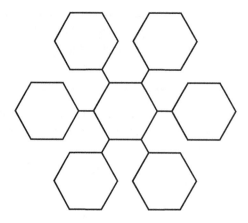

2 Fill in the digits from 1 to 7 in the blocks on the right, without repeating any to make this sum correct.

3 The triangle below must contain all the digits from 1 to 9 and each side must add up to 20. Fill in the missing numbers.

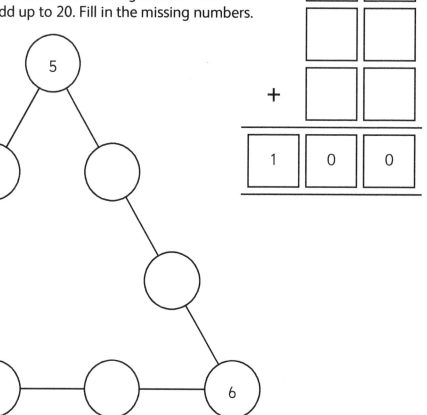

More Problems

1 Using exactly three straight lines, join these dots.

2 Draw exactly three straight lines on this diagram to separate each star from every other star.

3 Maria has arranged eight square napkins on the table in this design. The napkins overlap and only the top one (1) can be seen completely. Cut out 8 squares of paper to model the napkins. Arrange them to make the same pattern. Check your partner's arrangement to make sure it is correct.

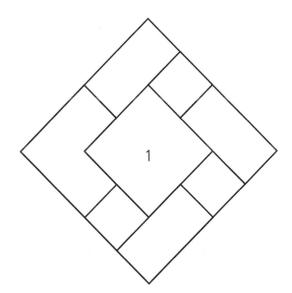

Expressions and Variables

1 Check these answers. Write TRUE or FALSE.

a $x + 14 - 30$, so $x = 16$ _____

b $y \div 3 = 18$, so $y = 6$ _____

c $100 \div x + 2 = 12$, so $x = 12$ _____

d $(x + 4) \times 3 = 70$, so $x = 3$ _____

2 Write a letter from the left to match the expression on the right.

a The sum of m and n multiplied by 3 _____ $(3 + m) \times n$

b The sum of three times m plus n _____ $3 \times m + n$

c The product of m and n less 3 _____ $m + m + n$

d The sum of 3 and m with the result multiplied by n _____ $(m + n) \times 3$

e The product of m squared and n _____ $2 \times m + 2 \times n$

f The sum of m and n, divided by 3 _____ $m \times 3 \div n$

g Twice m plus n _____ $m \times n - 3$

h m multiplied by 3 and then divided by n _____ $2 \times m \div n$

i The sum of m plus m plus n _____ $m \times m \times n$

j The quotient of twice m and n _____ $(m + n) \div 3$

k The sum of double m and double n _____ $2 \times m + n$

l m times n decreased by n _____ $m \times n - n$

Topic 16 Length

Estimate and Measure

1 For each animal below:
- estimate the length of its body in mm and write it in the table
- measure the length in mm and write it in the table
- calculate the difference between your estimate and the measured length.

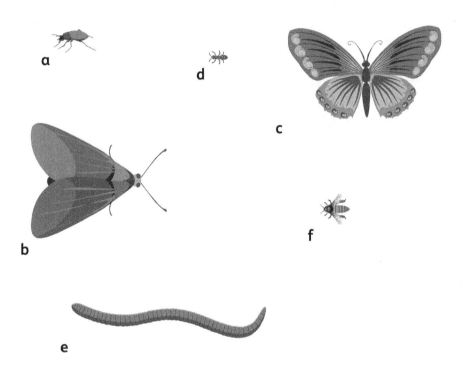

	My Estimate	Actual Measurement	Difference
a			
b			
c			
d			
e			
f			

2 Fill in this crossword grid about the different units of measure.

CLUES

Across

2 Usain Bolt holds the record for running this distance, which is also equal to ten dekametres!

6 About the thickness of a piece of very heavy cardboard.

7 Three of these would make the length of a normal school ruler.

Down

1 About the height of a tall palm tree; the unit that is equal to 10 metres.

3 The unit you would use to measure the distance between two villages.

4 The unit that describes the distance between your outstretched hands; equal to 100 cm.

5 The unit that is about the thickness of your finger.

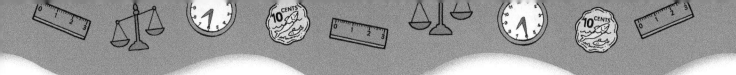

Find Me Game

Your teacher will give you cards with different lengths.

Work in pairs. Take turns to do the following.

- Pick a card.
- Find an object that you estimate has the same length that is on the card.
- Measure the object with a tape measure or ruler.
- Record the actual length of the object.
- Calculate the difference between the two measurements.

Continue until you have used all the cards in your set.

Length on Card	Object	Actual Measured Length	Difference

Which object's length did you estimate most accurately?

_____

Which object's length did you estimate least accurately?

_____

Using Decimals

1 Use a metre stick to help you work these out.

    a   One-tenth of a metre = 0.1 m = _____ dm

          Two-tenths of a metre = 0.2 m = _____ dm

          Three-tenths of a metre = _____ m = _____ dm

          Half a metre = _____ = _____ dm

          0.7 m = _____ dm

    b   One-hundredth of a metre = 0.01 m = _____ cm

          Two-hundredths of a metre = 0.02 m = _____ cm

          Three-hundredths of a metre = 0.03 m = _____ cm

          0.13 m = _____ cm

          0.75 m = _____ cm

2 Use your metre stick to help you. Complete the table.

Metres	Decimetres	Centimetres	Millimetres
0.002 m	0.02 dm	0.2 cm	2 mm
	0.06 dm	0.6 cm	
	0.3 dm	3 cm	30 mm
0.035 m		3.5 cm	35 mm
	0.82 dm	8.2 cm	
0.0255 m		25.5 cm	225 mm
0.45 m	4.5 dm		
	5.5 dm		
2 m			
2.085 m			

Topic 17 Multiplying and Dividing

Complete the flow charts.

a

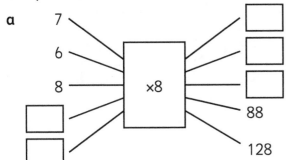

7
6
8

×8

88
128

b

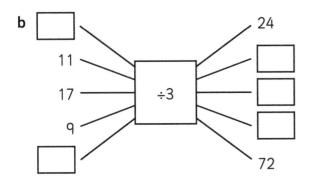

11
17
9

÷3

24
72

c

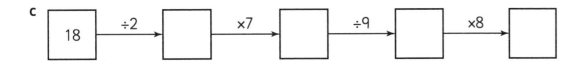

18 →÷2→ →×7→ →÷9→ →×8→

d

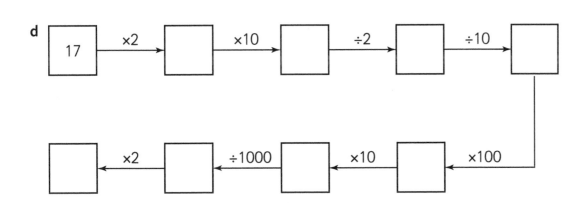

17 →×2→ →×10→ →÷2→ →÷10→

→×2→ ←÷1000← ←×10← ←×100←

Divisibility Rules

Test whether the numbers in the first column are divisible by 2, 3, 4, 5, 6, 8, 9 and 10.
Tick the columns to show your results.

Number	Divisible by ...							
	2	3	4	5	6	8	9	10
300								
245								
476								
389								
4 690								
4 307								
1 039								
2 340								
12 856								
4 526								
30 090								
12 345								
123 456								

Long Division

Find the missing numbers to complete these long divisions.

There is space below for working out each one.

a

				↓		↓	
			2		r		
2	3	4	6	7			
				↓			
			0				

b

			↓				
				1	r		
2	7	8	3	9			
		8	1	↓			
	–						
			2				

c

			↓		↓		
				8		r	
1	9	2	4	8	9		
	–	1	9	↓			
	–						
					9		
				3	8		

d

			↓	↓			
			1			r	
2	7	3	6	4	5		
	–	2	7	↓			
			9				
	–						
			1	3			
	–						

Topic 18 Order of Operations

Grouping Symbols

The answers to these calculations are correct.

Add grouping symbols to each one to show how you would work to get each answer.

a $9 \times 2 + 3 = 45$

b $16 - 7 \times 3 = 27$

c $20 \div 4 + 1 = 4$

d $20 \div 4 + 1 = 6$

e $18 + 9 \div 3 = 9$

f $64 \div 8 - 6 = 2$

g $4 + 8 \times 10 = 120$

h $19 - 9 \div 5 = 2$

i $5 \times 3 + 8 = 55$

j $10 \times 7 + 5 = 75$

k $40 \div 13 - 8 = 8$

l $10 \div 38 - 33 = 2$

m $20 \div 2 \div 2 = 5$

n $20 \div 2 \div 2 = 20$

o $4 + 2 \times 7 + 2 = 44$

p $4 + 2 \times 7 + 2 = 20$

q $56 - 37 - 56 - 47 = 10$

r $100 - 64 \div 6 = 6$

s $10 + 6 \div 4 + 4 = 2$

t $15 + 3 \times 0 = 15$

Order of Operations

1 The operation symbols have been erased in these calculations. Fill in the symbols to make each number sentence true.

a 5 ☐ 6 ☐ 8 = 22

b 38 ☐ 6 ☐ 2 ☐ 35

c 12 ☐ 4 ☐ 3 ☐ 3 = 16

d 32 ☐ 8 ☐ 3 = 12

2 Apply the order of operations rules to find each answer.
Then, add two sets of parentheses to each calculation to change it and work out the new answer.

a 3 × 4 + 3 × 5 + 3 × 6 = ☐   3 × 4 + 3 × 5 + 3 × 6 = ☐

b 2 × 8 + 2 × 9 + 2 × 7 = ☐   2 × 8 + 2 × 9 + 2 × 7 = ☐

c 8 × 4 + 8 × 2 + 8 × 1 = ☐   8 × 4 + 8 × 2 + 8 × 1 = ☐

d 4 × 5 + 4 × 3 + 4 × 2 = ☐   4 × 5 + 4 × 3 + 4 × 2 = ☐

Topic 19 Working with Time

Estimate, Compare and Measure Time

1 What time of day do you usually . . .

a get up in the morning? _____

b start school? _____

c finish school? _____

d play with friends? _____

e eat dinner? _____

f go to bed? _____

2 When does it happen? Write each date.

a My birthday _____

b Birthday of my friend _____

c Christmas Day _____

d New Year's Day _____

3 Think about things you did yesterday during the day and during the night. Write what you were doing at each of the following times:

a 1:00 a.m. _____

b 1:30 a.m. _____

c 6:30 p.m. _____

d 3:00 p.m. _____

Units of Time

1 Match each term in the first column to the best definition in the second column. Write a letter on the line provided.

Term		Definition
a Second	_____	24 hours
b Week	_____	12 months
c Minute	_____	10 years
d Century	_____	$\frac{1}{60}$ of a minute
e Day	_____	366 days
f Decade	_____	7 days
g Leap year	_____	60 seconds
h Year	_____	100 years

2 Complete the chart. Draw lines to show how long it takes.

	Seconds	Minutes	Hours	Days	Weeks	Months	Years	Decades
Make your bed								
Grow into a very old man or woman								
Brush your teeth								
Bake some cookies								
Smile								
Play a game of cricket								
Become an adult								
Grow an oak tree from an acorn								
Grow some radishes from seeds								

3 Write each date using SI format.

   a March 30th 1982 _____

   b February 5th 2002 _____

   c 18 October 1955 _____

4 An online article has the date 01/02/2012.
Write the date in words if this date is:

   a British format _____

   b American format _____

5 Write the times shown on these clocks.

a

   _____

b

   _____

c

   _____

d

   _____

e

   _____

f

   _____

6 Draw the hands on the clocks to show these times.

a half past 3

b quarter past 9

c quarter to 4

d 25 past 1

7 Draw the hands on the first clocks to show these times. Draw hands on the second clocks to show the time a quarter of an hour earlier.

a 25 to 12

b 20 to 11

c 5 o'clock

d half past six

Digital and 24-Hour Time

Complete the table to give the times in three different ways:

- in words
- in 12-hour notation (a.m. and p.m. time)
- in 24-hour notation.

Remember that 24-hour notation always has four digits. The first two tell the hour and the second two tell the minutes.

In Words	12-Hour Time	24-Hour Time
Half past seven in the morning	7:30 a.m.	07:30
Six o'clock in the evening	6:00 p.m.	18:00
Quarter past ten in the evening	10:15 p.m.	22:15
Nine o'clock at night		
Quarter to eleven at night		
Twenty-five to three in the afternoon		
Five minutes before midnight		23:55
	12:05 a.m.	00:05
		14:35
	8:45 a.m.	
		19:55
	4:47 p.m.	

Topic 20 Calculating with Fractions and Decimals

Magic Squares

In each magic square, the sum of each row, column and diagonal is the same.

Work out the missing numbers.

a

	$9\frac{9}{10}$	$4\frac{4}{10}$
	$5\frac{1}{2}$	
$6\frac{3}{5}$		$8\frac{4}{5}$

b

$\frac{1}{12}$		
	$\frac{1}{8}$	
$\frac{3}{16}$		$\frac{1}{6}$

c

1.2		1.6
	1	
0.4		0.8

d

0.75		0.25
0.125	0.625	
		0.5

e

2.625		
4.8125		2.6125
3.25		

f

	0.09	0.1
0.07	0.11	
0.12		0.08

Add and Subtract Decimals

The mass of different stationery items is given below.
The empty pencil case has a mass of 33.53 grams.

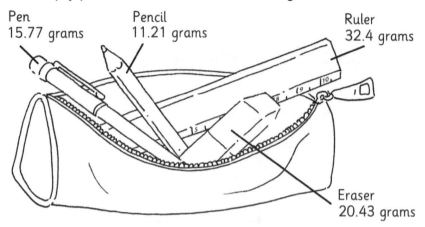

Pen
15.77 grams

Pencil
11.21 grams

Ruler
32.4 grams

Eraser
20.43 grams

The number of items in the case and the total mass is given below.

Work out what items the pencil case could contain in each case.

One _____ and one _____
Total mass: 60.51 g

Two _____ and one _____
Total mass: 76.38 g

One _____ and one _____
Total mass: 81.7 g

One _____ and one _____
Total mass: 65.17 g

Three _____, one _____ and one _____
Total mass: 115.33 g

One _____, two _____ and one _____.
Total mass: 96.71 g

Topic 21 Capacity and Volume

Measuring Volume

1 Calculate the volume of each box by adding blocks. Each block has a volume of 1 centimetre cubed.

a

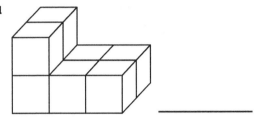

_____

b

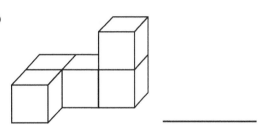

_____

c

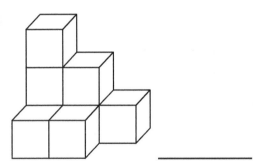

_____

d

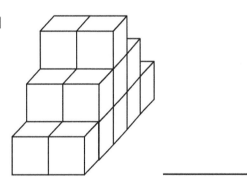

_____

e

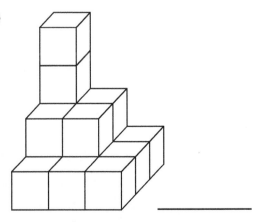

_____

2 Calculate the volume of the box in each sketch. Use the formula you have learned.

a

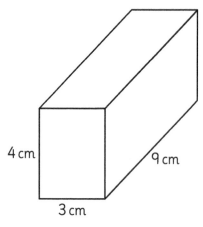

4 cm

9 cm

3 cm

b

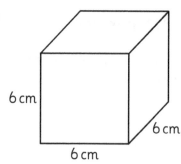

6 cm

6 cm

6 cm

c

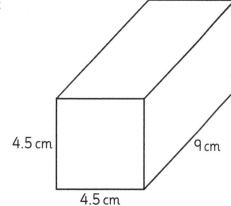

4.5 cm

9 cm

4.5 cm

d

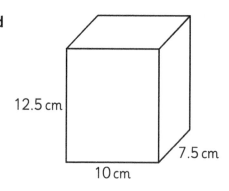

12.5 cm

7.5 cm

10 cm

3 Challenge! Give two different sets of dimensions that could make a box with a volume of 64 cm³.

_____

_____

Measuring Capacity

1 One jug like this holds a litre. The line at the top of the jug shows the litre mark. Write the approximate amount of water shown in each jug in millilitres.

a

_____

b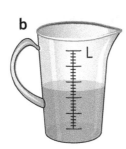

_____

c

_____

d

_____

2 How many times would you need to fill one of these containers and pour it into the jug in question 1 before you made a whole litre?

a

_____

b

_____

c

_____

d

_____

3 How many…

   a pints in a quart? _____

   b quarts in a gallon? _____

   c cups in a litre? _____

Topic 22 Transformations

Translation, Reflection and Rotation

1 Follow the instructions to complete the following drawings.

a Rotate the figure 180°.
To make it a bit easier, you can cut out a congruent stop sign similar to the one in the diagram, write on the word 'STOP' and then do the rotation with the cut-out shape.

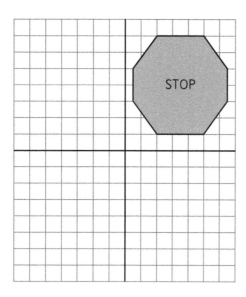

b Translate the following figure 3 blocks up and 5 blocks to the right.

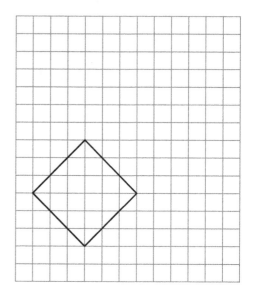

c Reflect the figure across the mirror line.

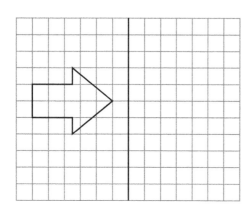

d Translate the figure by 3 blocks in any direction.

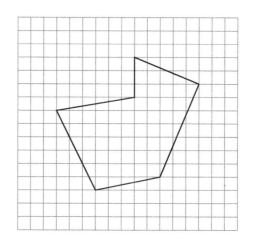

e Rotate the figure by 90° anti-clockwise.

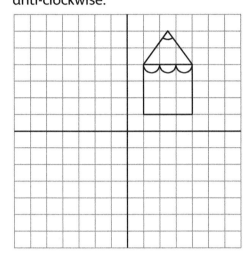

f Reflect the figure across the mirror line.

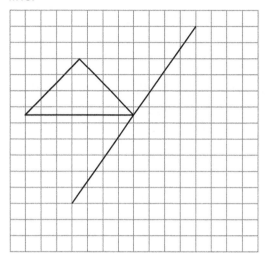

2 Use transformations to complete this symmetrical design around the lines of symmetry.

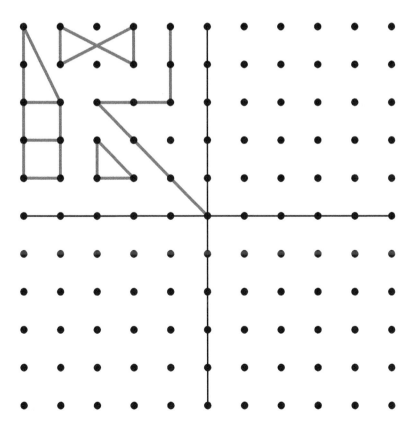

3 What does congruent mean?

_____

4 Draw a congruent and an incongruent figure in the labelled boxes for each of the shapes below.

Shape	Congruent	Not Congruent
a		
b		
c		
d		
e		

Topic 23 Perimeter and Area

Measuring Perimeter

1 Estimate then measure the perimeter of each shape.

Estimate: _____

Measurement: _____

Estimate: _____

Measurement: _____

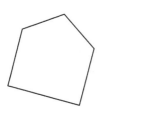

Estimate: _____

Measurement: _____

Estimate: _____

Measurement: _____

2 Calculate the perimeter of each shape.

a

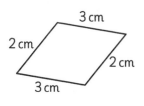

2 cm 3 cm 2 cm 3 cm

_____

b

3.5 km   3.5 km

2.7 km

_____

c

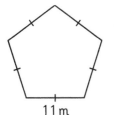

(Hint: the short lines through each side mean they are equal lengths)

11 m

_____

d

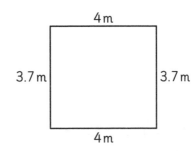

4 m

3.7 m   3.7 m

4 m

_____

Measuring Area

1 a Work out the perimeter and area of each shape.

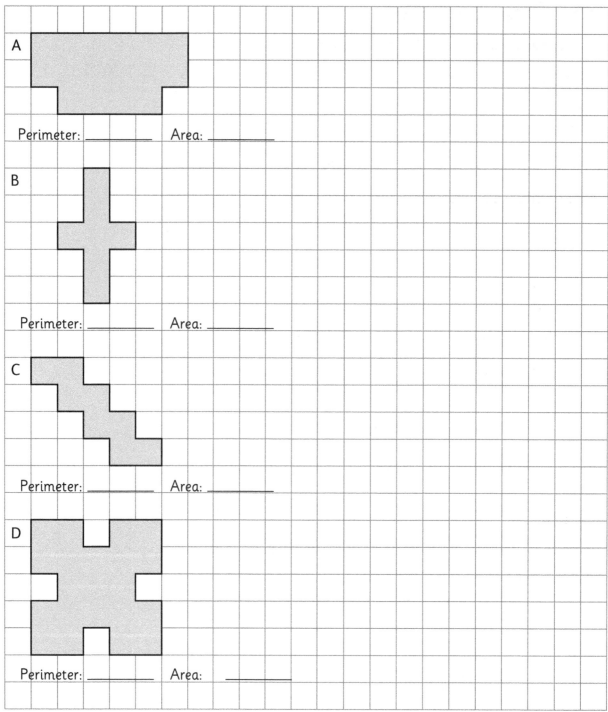

A

Perimeter: _____   Area: _____

B

Perimeter: _____   Area: _____

C

Perimeter: _____   Area: _____

D

Perimeter: _____   Area: _____

b Next to shapes A and B, draw another shape with the same area but different perimeter.

c Next to shapes C and D, draw another shape with the same perimeter but different area.

Topic 24 Probability

Chance, Outcomes and Probability

1 You will conduct a probability experiment with a partner. Make sure that you have a six-sided die.

   a First, answer these questions:

     What is the lowest number you can roll with one die? _____

     What is the highest number you can roll with one die? _____

     What is the most likely number you will roll with one die? _____

   b Complete the table to show the probability of rolling each of the numbers on a die.

	$P(1) = \dfrac{\text{Number of favourable outcomes}}{\text{Number of possible outcomes}} = \underline{\hspace{2cm}}$
	$P(2) = \dfrac{\text{Number of favourable outcomes}}{\text{Number of possible outcomes}} = \underline{\hspace{2cm}}$
	$P(3) = \dfrac{\text{Number of favourable outcomes}}{\text{Number of possible outcomes}} = \underline{\hspace{2cm}}$
	$P(4) = \dfrac{\text{Number of favourable outcomes}}{\text{Number of possible outcomes}} = \underline{\hspace{2cm}}$
	$P(5) = \dfrac{\text{Number of favourable outcomes}}{\text{Number of possible outcomes}} = \underline{\hspace{2cm}}$
	$P(6) = \dfrac{\text{Number of favourable outcomes}}{\text{Number of possible outcomes}} = \underline{\hspace{2cm}}$

What is the total of all the probabilities?

_____ + _____ + _____ + _____ + _____ + _____ = _____

Imagine you rolled the die 60 times. How many instances of each number can you expect?

_____

c Draw a bar graph to show the probability of rolling each number when you roll the die 60 times.

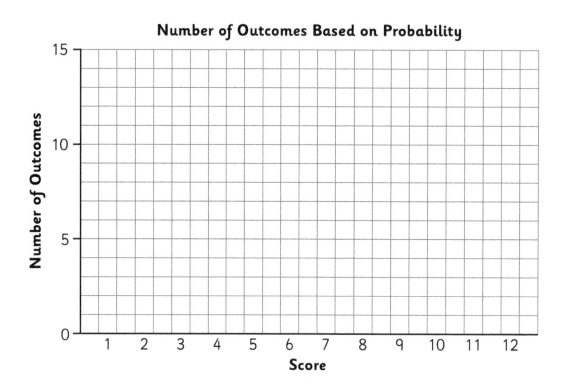

d Now discuss with your partner what would happen if you actually rolled the die 60 times. Would the die show each number exactly ten times? Why or why not? Write down your conclusion below.

_____

_____

e Let's do the experiment! Roll your die 60 times and use the tally table below to record your results.

Number	Tally	Frequency
1		
2		
3		
4		
5		
6		

f Draw a bar graph to illustrate your results after rolling the die 60 times.

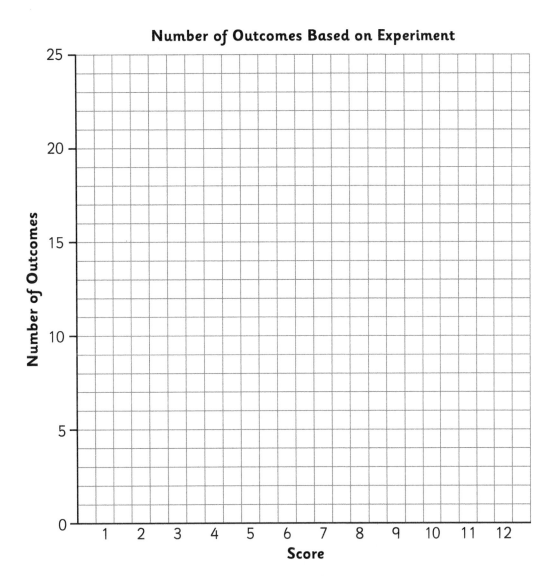

Number of Outcomes Based on Experiment

g How is the second bar graph different from your first bar graph?

_____

_____

h Compare your results to those of your classmates. Are they all the same?

_____

_____

i What can you conclude about the probability and the experimental outcomes of the event of rolling any number on a six-sided die?

_____

_____

_____